Better Homes and Gardens®

VEGETABLES
HERBS
AND FRUITS

Excerpted from Better Homes and Gardens® *STEP-BY-STEP SUCCESSFUL GARDENING*

© Copyright 1988 by Meredith Corporation, Des Moines, Iowa.
All Rights Reserved. Printed in the United States of America.
First Edition. First Printing.
ISBN: 0-696-01806-3

BETTER HOMES AND GARDENS® BOOKS

Editor: Gerald M. Knox
Art Director: Ernest Shelton
Managing Editor: David A. Kirchner
Editorial Project Managers: James D. Blume, Marsha Jahns,
   Rosanne Weber Mattson, Mary Helen Schiltz

Garden, Projects, and New Products Editor:
   Douglas A. Jimerson
Associate Editor: Jane Austin McKeon

Associate Art Directors: Linda Ford Vermie,
   Neoma Thomas, Randall Yontz
Assistant Art Directors: Lynda Haupert, Harijs Priekulis,
   Tom Wegner
Graphic Designers: Mary Schlueter Bendgen, Mike Burns,
   Brian Wignall
Art Production: Director: John Berg;
   Associate: Joe Heuer
   Office Manager: Michaela Lester

President, Book Group: Fred Stines
Vice President, General Manager: Jeramy Lanigan
Vice President, Retail Marketing: Jamie L. Martin
Vice President, Administrative Services: Rick Rundall

BETTER HOMES AND GARDENS® MAGAZINE
President, Magazine Group: James A. Autry
Vice President, Editorial Director: Doris Eby
Executive Director, Editorial Services: Duane L. Gregg

VEGETABLES, HERBS, AND FRUITS
Editorial Project Manager: Rosanne Weber Mattson
Graphic Designer: Brian Wignall
Electronic Text Processor: Paula Forest

CONTENTS

VEGETABLES IN THE LANDSCAPE

With some planning and a little patience, your vegetable garden can become the most eye-appealing element in your landscape. In fact, a neatly groomed vegetable garden can be just as beautiful as a rose garden or perennial border (and certainly tastes better).

HOW TO GET STARTED

■ Vegetables aren't fussy. Their needs are really pretty basic: lots of sunshine, fertile soil, moisture, and a kind hand now and then when weeds or insect pests try to take over. A vegetable garden needn't require a lot of work, either. With early planning and some laborsaving techniques like mulching, wide-row planting, and drip irrigation, you can relax once the rush of spring planting is over. A daily trip to the garden to check your crops' progress and to pull a few weeds is the only summer gardening chore you'll have.

When you're planning your garden, look for a site that gets at least 6 to 8 hours of direct sunlight a day. A level spot is best, but even a steep slope can be tamed with terraces cut into the hillside. Don't locate your garden under trees or near tall, established shrubs. These plants will keep your vegetables from getting sunshine, and their extensive root systems will rob your crops of nutrients and moisture.

SOIL PREPARATION

■ After you've selected your garden spot, it's time to improve soil quality. Even the best soils can stand improvement, so don't skimp when it comes to soil preparation. To get a good indication of your soil's needs, take soil samples and have them analyzed by your extension service, county agent, garden center, or nursery.

Clear the area of branches, stones, and other debris. Strip the sod if you're starting in a grassy area. Compost, rotting leaves, straw (be sure it's weed-free), and manure are all excellent soil amendments. Peat moss, bone meal, dehydrated manures, and cottonseed meal are a few other amendments you can use, depending on the results of your soil test. Till everything under and rake smooth the surface of the garden.

AUTOMATIC WATERING

■ To eliminate watering chores, you may want to install a drip irrigation system after your vegetables are up and growing. Drip systems come in a variety of styles, but they all work on the same principle. Water is delivered in small quantities under low pressure directly to where it does the most good—the root zones of the plants.

Raised beds (*left*) can enhance your home's landscape. You can learn a lot from demonstration gardens. The one at *right* is in St. Louis.

VEGETABLES IN THE LANDSCAPE

Good soil conditioning is the key to the success of the well-planned 40x100-foot seaside garden at *right*. In fall, the gardener spreads several inches of seaweed and rotted poultry manure over the top of the garden and allows it to decompose. In the early spring, all of this organic matter is turned under with a rotary tiller.

When the crops are up and growing, the soil around them is kept cool and moist with a thick mulch of salt hay. Every time a crop is harvested during the growing season, the soil is turned over before another crop is planted. Weeds, which rob the soil of nutrients and moisture, are pulled whenever they appear. Paths, too, are cultivated to keep weeds from getting a foothold.

The magnificent raised-bed garden at *right* is more than just a lush collection of vegetables and flowers. It's also a glorious spot for summer picnics and family get-togethers. The paths between the beds and the patio in the center of the garden are all maintenance-free because they're covered with black plastic topped with a thick layer of pea gravel. There are eight beds in all, four 8x50 feet and four 6x6 feet, which are contained in treated hemlock boards.

To keep the flowers in shape, they're fed a commercial slow-release 14-14-14 mixture of nitrogen, phosphorus, and potassium. Vegetables get a mix of seaweed powder and water, sprayed on their leaves at regular intervals.

For colonial gardeners (*opposite*), a vegetable garden was a necessity.

VEGETABLE GARDEN PLANNING

Once the grip of winter is broken, nothing can stop the force of spring—not even a procrastinating gardener who fails to put growing plans in order. In no time, the soil will be ready for preparation as the threat of frost fades. If proper plans aren't made beforehand, spring can turn into a race for time instead of a season of joy and expectation.

ARMCHAIR PLANNING
■ The first step is to sit down and list the vegetables you want to grow. Be sure to take into account the likes and dislikes of the rest of the family so you're not left with piles of unwanted produce. Then consider some of the space-saving and yield-boosting gardening techniques, such as companion planting, succession planting, intercropping, and second-cropping (see *Vegetable Techniques,* page 10). If this is your first garden, you may have to brave the cold to measure the actual space available for your vegetable crops.

Be as accurate as possible. It may require some extra calculations, but you can save considerable money and time by knowing exactly how much space you have to work with. Then transfer your measurements to a sheet of graph paper so that the sketch matches—to the inch—what you have outside. All that remains is to mark the location of the rows and the vegetables that will occupy them. To avoid midsummer gaps in your garden, plan to replace harvested vegetables with second crops. And also indicate what varieties you plan to squeeze in between rows of other vegetables.

WHEN TO PLANT
■ Proper timing is critical to successful gardening. There are long-season and short-season vegetables, and crops that do better in cooler spring temperatures or warmer summer temperatures.

Long-season crops are those that take the entire season to grow, flower, and produce mature fruit. Cantaloupe, watermelon, winter squash, potato, tomato, corn, cucumber, pumpkin, pepper, and eggplant fall into this category. Short-season vegetables practically explode out of the ground, which means you can plant several times within a single season. Radishes, beans, lettuces, beets, and carrots can be sown and harvested before the season is half over.

Cool-season crops can be planted as early in the spring as you can work the soil. These crops include radish, asparagus, beet, broccoli, Brussels sprouts, cabbage, carrot, cauliflower, celery, Swiss chard, chive, leek, lettuce, onion, parsley, parsnip, pea, potato, rhubarb, spinach, and turnip.

Warm-season crops must be planted after the threat of frost has passed. Those that need warmer temperatures include beans (green and lima), corn, cucumber, eggplant, cantaloupe, pepper, pumpkin, summer squash, winter squash, tomato, and watermelon.

KEEPING RECORDS
■ Because your memory may fade over winter, it's a good idea to keep written records of what you planted and how you did it. Simply note the crops and varieties you grew, and when and how much you planted and harvested. Re-member to leave ample space for your own comments. Keeping track of the varieties you grow will give you the chance to experiment with different varieties. The record of planting dates will give you a clear idea of when to plant next time around.

RAISED-BED GARDENING
■ One popular method that increases yields in small gardens is raised-bed gardening. This system (*right*) works on the principle that deeply dug, fertile soil will allow you to plant more crops in a smaller amount of space.

A raised-bed garden can be any length, but should be no more than five feet wide to let you reach in from either side. First, remove topsoil by digging trenches about 12 inches wide and 12 inches deep. Loosen the subsoil, then mound the topsoil on top. Your finished bed should be four to six inches higher than ground level. Taper sides so rainfall won't wash soil away. Plant seeds in wide bands instead of in individual rows.

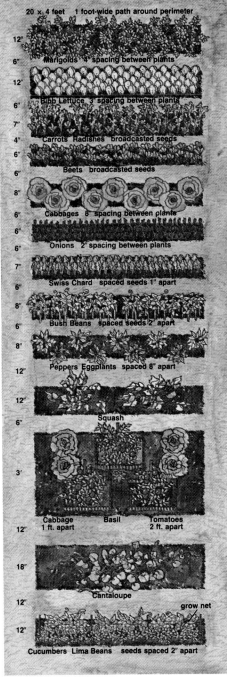

20 x 4 feet   1 foot-wide path around perimeter

12″

6″   Marigolds 4″ spacing between plants

12″

6″   Bibb Lettuce 3″ spacing between plants

7″

4″   Carrots Radishes broadcasted seeds

6″

6″   Beets broadcasted seeds

8″

6″   Cabbages 8″ spacing between plants

6″

6″   Onions 2″ spacing between plants

7″

6″   Swiss Chard spaced seeds 1″ apart

8″

6″   Bush Beans spaced seeds 2″ apart

8″

12″   Peppers Eggplants spaced 8″ apart

12″

6″   Squash

3′

12″   Cabbage   Basil   Tomatoes
       1 ft. apart   2 ft. apart

18″

12″   Cantaloupe   grow net

12″

Cucumbers Lima Beans   seeds spaced 2″ apart

VEGETABLE TECHNIQUES

DOUBLE CROPPING

■ Even if you have an acre of garden space, it's still wise to double up plantings. You'll get twice the harvest with little increase in work. Onions to be used as scallions can be set among cabbage, broccoli, cauliflower, or Brussels sprouts plants. Radishes can be sown with slower germinating carrots. Radishes sprout quickly and prevent weed growth. When radishes are ready to harvest, the carrots can take over. Beets and broccoli also are a good combination. Beets grow rapidly in spring while broccoli is getting established.

The onions and carrots in the garden at *right* can be harvested while still small, leaving space for the lettuce. Or, harvest the carrots and lettuce and allow the onions to develop fully.

COVER CROPPING

■ Cover crops add vital nutrients to the soil, improve drainage, and keep weed growth choked out on fallow areas. Annual rye grass, one of the most popular cover crops, is usually planted in the fall after all the vegetables are harvested. The grass grows quickly during wet, cool fall weather and dies down when the soil freezes. In spring, till the grass under before planting time.

Cover crops are useful in midsummer, too. After cool-weather crops die, cover crops like buckwheat (*right*) can be sown to fill in empty spaces. Or, use vegetables like peas and beans as cover crops in the spring and summer and get a harvest before you till the plants under to add organic matter.

VERTICAL GARDENING

■ If you're short on growing room, try vertical gardening. A sunny section of chain-link or woven-wire fence can support a big crop of vegetables. Climbing peas, pole beans, melons, pumpkins, and squash will grow on vertical supports. Melons and squash fruits need sling supports to keep the weight from pulling vines off the fence. Improvise slings of net and tie to the fence.

Another space-saving trick is to grow vine crops on trellises (*left*) made of lath stakes or bamboo poles. Plant seeds at the base and get ready to harvest a triple-size crop. If you have a surplus of shrub prunings, pile them up and let beans ramble over the stack. If you have a long growing season, plant an early pea crop and a late bean crop.

WIDE-ROW GARDENING

■ One of the best ways to increase yield without tilling up more space is to garden in wide rows. Wide-row gardening works on the principle that plants sown close together (*left*) produce up to four times as much harvest as the same area planted single file. Individual plants might not produce as much as when they're spaced according to seed pack instructions, but on the whole, volume of production will be much greater.

Wide-row gardening eliminates a lot of weeding, because vegetables grow thickly to choke out weeds. When you grow cool-weather crops like lettuce, spinach, and peas in the wide-row system, you'll also be extending your harvest, because the plants will shade each other during hot weather.

SMALL-SPACE GARDENS

Truly successful gardeners plan their vegetable gardens to get two or three major harvests in one season. When one crop matures, another is planted. And, whenever possible, two crops with different maturity dates are planted together for a staggered harvest.

PLANT EVERY SQUARE INCH

■ To ensure continuous production, keep garden space filled. The garden at *right* is a perfect example of multiple-harvest planning. The large 10x16-foot bed is planted with cool-season crops in early April—spinach, lettuce, beets, radishes, carrots, onions, and peas. By late May, most of these crops have been harvested and the warm-weather crops planted, including beans, cucumbers, cantaloupes, and more carrots. Cool-season crops that remain have warm-weather crops planted around them.

The 5x8 bed is planted with long-maturing crops. Here, vegetables like eggplant, broccoli, cabbage, and corn grow undisturbed until harvest. Even these long-season vegetables are replaced in late summer with lettuce, radish, and kale for late-fall harvest.

PREPARE SOIL CAREFULLY

■ You'll need to prepare your soil well to keep your garden in constant production, as shown at *right*. First, both beds were staked out and 1x8 side-boards installed. Then, the beds were dug to a depth of 4 inches and layers of compost, manure, sawdust, and soil were added until the soil level reached the top of the boards. Lengths of 2x4s set over the beds let this gardener tend the crops without walking on the soil.

GETTING STARTED

A seed holds an incredible life force. When conditions are right, the seed bursts, sending forth an embryo root and stem. Each time, the same thing happens with mind-boggling regularity. But the key to the process is to give the right seed the right conditions—which is the gardener's job.

Good germination needs moisture and warmth. Ideally, the temperature should be between 70 and 80 degrees Fahrenheit. Moisture is necessary to soften hard seed shells and provide nutrients for developing roots.

Naturally, soil conditions are important for germination, too. After you've turned and fortified the soil with lime, fertilizer, and organic matter to improve its content, mark the exact location of the row. Place small stakes at each end, and stretch twine between them. Then, using the twine as a guide, dig a furrow with the corner of a hoe blade (or the handle end if a shallow furrow is needed). The idea is to plant seed at the right depth. The deeper you go, the cooler the soil temperature; but the shallower the furrow, the drier the soil. A good rule of thumb is to make sure to plant seed at a depth equal to four times its diameter.

SOWING SEEDS OUTDOORS

■ If seeds are small, take a pinch between thumb and index finger and sow by rubbing fingers together over the furrow. Larger seeds can be planted individually. Cover seed with soil and tamp gently with your hand or the back of your hoe blade. Remove twine and mark the row with an empty seed packet or plant label. Water gently but thoroughly; if a dry spell occurs, keep soil moist by placing a thin mulch of dried lawn clippings or partially rotted compost over the row.

PLANTING SEEDLINGS

■ When setting out young seedlings, mark rows with twine and dig holes at recommended intervals. Fill holes with water, allow them to drain, and set in plants after placing foil or paper collars around stems to prevent cutworm damage. Then fill in soil, being sure to eliminate unwanted air spaces, and tamp gently. If plants appear to be suffering from the intense sun, shade can be supplied by placing shields made of brush or small boards next to each plant. Water frequently.

When you start plants from seeds outdoors, it may be necessary to thin out some plants. Use scissors or shears and cut the stem at ground level. Pulling can damage neighboring seedlings.

CABBAGE
EARLY

EXTENDING THE SEASON

Growing vegetables outdoors year-round is easy, even in northern climates, when you extend your growing season with a cloche or cold frame. These structures are designed to take full advantage of the sun's rays so that you can start gardening earlier in the spring and continue to produce harvests well after the first frost. In some cases, you may even be able to grow crops in midwinter when the rest of the garden is covered with snow.

For example, the cold frame *opposite* has a double-glazed fiberglass top that's curved to take full advantage of the sun's rays and allow plenty of growing room for the crops. A 55-gallon drum, painted black and filled with water, absorbs heat from the sun and helps keep temperatures stable.

Wall O' Water caps let you protect individual plants. Filled with water, tubes capture heat from the sun during the day and stay warm as evening temperatures drop.

This portable cold frame has a 3x4-foot frame made of 1x2 redwood lumber. Clear plastic sheeting forms the walls. A plastic cover can be unrolled over the top.

Newly germinated celery plants are sheltered by panes of glass held together with aluminum clips. Celery is slow to germinate; this method accelerates growth by warming the soil.

Start crops like broccoli and cauliflower under row-long plastic tents. Each tent has two sheets of thick plastic—one clear and one black. The black plastic absorbs heat and helps keep the soil warm.

This cold frame is made from railroad ties and old windows. When the temperature inside rises, the window is propped up. When the temperature cools, the window is closed.

RAISED-BED GARDENS

One of the best raised-bed gardens in the country belongs to Peter and Sylvia Chan. To improve their soil and to beef up vegetable production, the Chans dug a system of raised beds similar to those used in southern China. Their garden (*opposite*) measures 50x25 feet and consists of ten 25-foot-long raised beds. The beds are four feet wide and separated by foot-wide paths.

The Chans use a small grow-light unit set up in their study to help start vegetables like broccoli, lettuce, beans, kohlrabi, and spinach. They sow seeds three to five weeks before planting in the garden. When the seedlings are about an inch high, the tiny plants are moved into individual two-inch pots. Once established in the pots, the plants are moved outdoors into a portable cold frame to harden off.

After a week or so in the cold frames, the seedlings are moved into their permanent locations in the garden. Peter adds a little dehydrated chicken manure to the bottom of each hole and tops the fertilizer with a thin layer of soil to prevent damage to the seedling's roots.

Later, as the crops grow, Peter feeds them once a week with a diluted solution of manure tea. Crops like peas and beans are sown directly into the beds with a little dehydrated chicken manure in the bottom of each furrow.

GETTING STARTED

■ Soil preparation is the key to this garden's success. First, the location of each bed is marked off with twine. Then, a sharp spade is used to turn the soil in the bed to a depth of ten or 12 inches. As the soil is turned, rocks, weeds, grass, and other debris are removed. After digging, the level of the bed is higher than the area around it because the soil is in a well-worked state; no new soil has been added. The sides of the beds are tapered so heavy rainfall won't wash the soil away.

All this might sound like a lot of work, but remember that once a raised bed is established, it's permanent. You needn't redig or till the garden every spring, as you must do with a flat garden. Raised beds also dry out more quickly in the spring, giving you a head start on the season. Raised beds do, however, require more frequent watering than traditional flat gardens.

In the fall, Peter and Sylvia dig a trench down the center of each raised bed. They then spread compost and organic matter from their compost bin and refill the trenches with soil. In the spring the Chans rake the beds smooth.

To save space, the Chans train pole beans and peas up lightweight wooden supports made from lengths of 1x2 and 2x2 lumber. Each support is portable and fits over an entire raised bed.

CARE AND MAINTENANCE

The quality and quantity of the vegetables you grow depends on whether the soil offers plants a good root environment and good nutrition. For the plant to thrive, the roots must penetrate the soil easily and draw adequate food.

If you already have a vegetable garden, your soil is probably in good shape. However, if you are breaking new ground, your soil may be less than ideal the first year.

Turn or till the soil a full-spade depth. Add all the organic material you can find and turn it under. Leaves, grass clippings, well-rotted manure, straw, compost, and leafy kitchen scraps are good organic soil improvers.

If your soil is acidic, scatter lime over the surface before digging. For heavy clay soil, try gypsum to modify the soil and make it easier to work.

After you've dug the soil and raked it roughly level, sprinkle the surface with a vegetable garden fertilizer. Be sure to follow the directions on the bag. After scattering fertilizer, rake it into the top two inches of soil.

The soil will be in the best condition for planting if you turn it in late fall—and you'll also lighten your gardening chores for the following spring.

WATERING
■ One inch of gentle rain each week during the growing season is every gardener's dream. However, this rarely happens. Don't be frugal when watering early in the season because roots are shallow and the plants will not develop optimally and bear fruit if the soil is allowed to dry out.

MULCHING
■ To save work, time, and money, mulch your garden in late spring. This protective cover insulates the soil against the heat of the summer sun, protects it from the drying winds, and all but eliminates weeds. Once the mulch is applied, you don't have to hoe between the rows, and, during most growing seasons, you have to water about one-third as often. And, as the organic mulches decay, they improve the soil. Wait to mulch until after the soil has warmed and before the weeds have started growing.

You can choose between organic mulches (straw, peat moss, sawdust, dry manure, and bark chips) and inorganic mulches (paper and polyethylene film). Inorganics work for vining crops like melons and squash.

A tomato plant can send out roots from any part of its stem. For extra-leggy plants, bury the entire stem lengthwise in a trench under the soil surface.

Cutworms feed on seedling stems. To combat these pests, wrap the bottom inch or two with a newspaper or foil collar extending just below the soil. Remove collar after plants are established.

For continuous harvests, vegetables like beans and peas should be picked often; during the peak of the season, you may harvest daily. Pods should be picked before seeds swell to cause visible bulges.

To blanch cauliflower heads, use twine or a rubber band to tie plant leaves over the top of developing head. This will shade the head, keeping it compact and white.

Some cool-weather crops like lettuce and spinach will bolt (go to seed) during hot weather. When this happens, pull plants and throw them on your compost pile. Choose more heat-resistant varieties.

Harvest broccoli when heads are tight and compact, usually when they're about 5 inches in diameter. If you wait too long, the head will produce small yellow flowers and be inedible. After you cut the main head, leave the plant intact. In a few weeks, smaller heads (*above*) will develop from side shoots along the plant's stem. Cut these smaller heads when they're about an inch or two across.

INSECTS AND DISEASES

INSECT/DISEASE	DESCRIPTION	CONTROL
Aphids	Aphids, or plant lice, are small, sucking insects in green, red, or brown, which attack beans, cabbages, peas, peppers, potatoes, and tomatoes. Plants attacked by aphids lose their vigor. Plant growth becomes stunted and often distorted. Aphids also can carry mosaic and other viral diseases.	Aphids can be washed off with a heavy stream of water or can be squashed by hand. Chemical controls, applied as soon as aphids appear, work swiftly and effectively.
Beetles	Beetles are visible, six-legged insects that feed on foliage and fruit. Beetle types include the asparagus, bean, striped or spotted cucumber (cucumbers, melon, squash), flea (tomatoes, turnips, radishes, potatoes, eggplant), and Colorado potato beetles. In many cases the larvae are also destructive.	Some can be picked by hand; otherwise use chemical controls made specifically for vegetables. Grub control is effective. Japanese beetles can be controlled with biological controls.
Borers	Corn borers get into the ears and eat the kernels. These same insects also attack peppers. Squash vine borers get into the stems of squash, melons, and cucumbers, causing the vines to wilt or die.	You can sometimes remove borers by hand, but it's hard to completely eliminate them this way. Consider controlling them chemically; be sure to follow directions for vegetables.
Caterpillars	These butterfly and moth larvae come in many forms, including cabbage looper, cabbageworm, corn earworm, fall armyworm, and tomato hornworm. They make their diets on leaves and fruit. Caterpillars tend to blend in with the background, but you can usually spot them if you look closely.	Some can be hand-picked; many are controlled biologically with a spray of *Bacillus thuringiensis* (sold as B.t.). Chemical sprays made for vegetables are also effective.
Cutworms	A cutworm is a type of caterpillar that emerges from the ground at night and eats young seedlings of tomatoes, cabbage, peppers, beans, and corn, cutting them off at the soil line and killing them.	For the most effective control, place a metal container or milk carton around each new plant so the cutworms cannot reach the stems.

INSECT/DISEASE	DESCRIPTION	CONTROL
Fungus diseases	Spots or marks on foliage, yellowing, leaf drop, stunted growth, wilting, dieback, distortion, or coatings on foliage can be caused by anthracnose, yellows, clubroot, scab, wilts (fusarium or verticillium), smut, blight, or mildew. These diseases are widespread in humid areas, and in wet, poorly drained soils.	Improve soil drainage. Plant only resistant varieties where available; rotate crops over a several-year period. Treat seed with fungicide before planting. Get rid of weeds. Avoid over-watering, crowding, and poor air circulation.
Leafhoppers	Leafhoppers are small insects, typically wedge-shape and green. They suck juice from beans, carrots, potatoes, lettuce, squash, and tomatoes. The plants lose vigor, turn yellow, and are more prone to diseases that the insects may be carrying.	Use a mulch of aluminum foil to keep the leafhoppers from jumping around. A chemical spray applied to the foliage will also control them.
Maggots	Maggots are fly larvae that generally attack roots or plant parts below ground level. They include seed-corn (corn and beans), cabbage root (cabbages, radishes, turnips), and onion maggots, and leaf-miners that tunnel beneath leaf surfaces of spinach, peppers, tomatoes, cucumbers, melons, or beets.	Use a mulch of black plastic or similar material to keep the maggots from boring below the soil line. Insect control applied to the soil at planting time also works.
Mosaic	This is a virus disease that causes stunting, mottling, distortion, and discoloration of leaves and stunting and streaking of fruit on beans, cabbages, cucumbers, melons, squash, peas, peppers, or spinach.	Plant resistant varieties where available. Keep weeds out. Control aphids, which cause the spread of mosaic.
Slugs and snails	When you see a slimy trail and you notice leaves disappear overnight, especially on young seedlings, you know you have slugs or snails. They also bore into mature, fleshy fruit, such as tomatoes, eggplant, peppers, melons, and squash.	Baits are available to use in the vegetable garden. You also can control slugs with beer or grapefruit halves placed around the garden. Hand-pick at twilight or after dark.

HERBS

Today's gardeners enjoy herbs for their fragrance, flavor, and rich ancestry. Many gardeners devote space to herbs simply because they are easy to grow. Herbs will thrive in places where other plants generally won't, demanding only sun or partial sun, and well-drained soil.

Feature your favorites in a separate herb garden, either formally to imitate a stately royal garden, or informally to create a natural thicket in your backyard. Or tuck a few herbs into existing gardens to provide a lush backdrop for flowers or a border around vegetables.

Herbs traditionally are planted in geometric shapes, with plants grouped according to flavor or fragrance. One of the simplest designs consists of four beds with intersecting paths. The design can be modified by enlarging the beds and adding more paths.

Popular cooking herbs include savory, basil, spearmint, chives, dill, oregano, parsley, and sage. Mint, scented geranium, thyme, lavender, and rosemary are favorite fragrant types. Cold-sensitive herbs can be grown in pots, then taken indoors for the winter.

GALLERY OF HERBS

The term herb covers a wide range of plants grown for culinary purposes, fragrance, medicine, cosmetics, or dyes. Herbs can be annual, biennial, or perennial. Collectively, they have similar growing requirements. The annual herb family includes basil, chervil, coriander, dill, borage, marjoram, and summer savory. There is only one well-recognized biennial: parsley. Perennials include oregano, sage, lavender, tansy, rosemary, tarragon, mint, and thyme. However, many will survive only in warm areas, so many gardeners consider them annuals. Annuals such as dill easily reseed, which makes them appear to be perennial.

Most herbs (with the exception of French tarragon and true oregano) are easily grown from seed. Try division or cuttings for others.

BASIL, SWEET
Ocimum basilicum

This annual herb is one of the most ornamental, with broad, clove-scented leaves of green or purple and a spike of white, purple, or pink flowers. The plant grows 15 to 24 inches tall and can be used outside or indoors as well.

BORAGE
Borago officinalis

A decorative annual herb filled with drooping clusters of purple or blue star-shaped flowers all summer. The 2-foot-tall plants have hairy, coarse leaves. Leaves and flowers are used as flavoring. Borage is hard to transplant.

OREGANO
Origanum vulgare

Some gardeners call oregano wild marjoram. This tender, shrubby perennial has aromatic foliage that grows 2½ feet tall. Sow seeds in partial shade after frost. Pinch back to make it bushy.

PARSLEY
Petroselinum crispum

Parsley is actually a biennial best grown as an annual. Leaves are dark green, divided, and curled. Plants are 12 to 18 inches tall. If grown for a second year, harvest leaves before flowers bloom.

SAGE
Salvia officinalis

Sage is a semishrubby, 2- to 2½-foot perennial with white, oblong, woolly leaves and flower spikes in violet-blue. Cut back occasionally to make it bushy. Hang stems to dry after cutting.

DILL
Anethum graveolens

Dill is an annual herb that reaches 2 to 3 feet tall, with finely divided, light green leaves and flattened clusters of tiny yellow flowers in mid-spring. The blooms produce small flat seeds used for flavoring and pickling.

LAVENDER
Lavandula species

Lavender is an aromatic perennial, deciduous or semievergreen, with a number of branches topped with whorled spikes of blue-violet flowers in early summer. Foliage is greenish gray. To keep compact, prune back after bloom.

MARJORAM
Origanum majorana

This annual, 2-foot plant has oval leaves that are velvety and aromatic and tiny white or orchid flowers in midsummer. Pick leaves anytime for fresh use and just before blooming for drying. Start seeds indoors 8 to 10 weeks early.

TANSY
Tanacetum vulgare

A hardy perennial, tansy grows 36 inches tall with finely cut leaves and flat clusters of small yellow flowers. This is a vigorous grower which can tend to get weedy. Divide plants in early spring and set 2 feet apart.

ROSEMARY
Rosmarinus officinalis

This 3-foot plant is hardy in warm parts of the country. Tiny, pale blue flowers bloom in winter or in early spring over aromatic grayish green, needlelike foliage. In the north, pot plants in fall; place in a sunny window during winter.

THYME
Thymus species

Thyme is a perennial ground cover, with tiny, gray-green leaves and a cluster of small, violet-blue flowers in spring or summer. There is also a winter thyme, which is a taller, shrubbier plant with rose blooms.

CARE AND MAINTENANCE

Herbs are easy to grow if you meet a few basic requirements. Most herbs do best in the sun. Herbs' essential oils, which account for their flavor and fragrance, are produced in the greatest quantities when the herbs receive 6 to 8 hours of sunshine a day.

Most herbs will thrive in any good garden loam. A soil pH of neutral to slightly alkaline, which can be corrected with lime, is best. Well-drained soil is essential; improve poor-draining soils with sand and organic material. Once established, most herbs are extremely hardy. Fertilize heavily harvested herbs such as basil, chives, and parsley. Too much water and/or fertilizer will produce lush foliage but low oil content and, therefore, you'll harvest little flavor or fragrance.

Weeds are undoubtedly an herb's worst enemy. If not controlled from the start, weeds will choke out the young plants. A friable soil prior to planting will discourage competition at the start. For extra protection, spread two or three inches of mulch around your herbs after they are established.

The leaves or tops of most herbs should be harvested when fresh and green—just before full bloom. Cut plants after morning dew evaporates, and hang upside down.

Another drying method is to spread the herbs on wire racks until they're completely dry. Crumble and store the herbs in airtight containers.

FRUIT IN THE LANDSCAPE

Fruit- and berry-producing plants are more than natural food factories. They also add distinctive colors, shapes, and fragrances to a landscape plan.

The backyard apple orchard *below* is awash with blooms in the spring, offers shady relief in the summer, and bears bushels of fresh produce in late summer and early fall. Even if you don't have plenty of open-yard space, you can plant several dwarf fruit trees.

Fruit trees don't always have to be planted in the regimented rows of an orchard to bear fruit. You can use pears, peaches, or dwarf apple varieties as accent plantings near your house. Because fruit trees bloom in the spring, they are perfect mates for foundation and ornamental plants. Keep in mind that some fruits will need companion trees for proper pollination.

POT UP MINIATURE TREES FOR SMALL SPACES
■ Miniature fruit trees are the answer for small yards. These waist-high wonders rarely grow over 6 feet tall, yet they bear only a year or two after planting. Just three years old, the tiny Bonanza peach *opposite* already produces big crops of luscious fruit.

Also called genetic dwarfs, miniature fruit trees are available in peach, nectarine, cherry, apple, apricot, pear, plum, and almond varieties. But don't let the terms "miniature" or "genetic dwarf" fool you. These short, stocky trees develop full-size fruit that's just as delicious as that from taller trees.

Along with being productive and good looking, container-grown miniature fruit trees also are portable. Use them to enhance a bare deck or patio, flank a front door, or team with a collection of potted flowers. For easy moving, install casters on the bottom of the container at planting time. If you live in the Deep South or on the West Coast (wherever winter temperatures remain above 25 degrees F.), you can grow miniature fruit trees in the ground.

Even if they didn't produce fruit, miniature fruit trees would still be a gardener's dream. These stocky charmers never outgrow their positions in the landscape. Pruning chores are minimal and the plants' compact growth habit makes insect and disease control easy.

In the North, container-grown miniature fruit trees need protection during the winter months. After nighttime temperatures begin to drop below 25 degrees F., move your trees into an unheated garage, shed, or porch. Water the container thoroughly and mulch it with a thick layer of leaves or straw.

GROW GRAPES FOR SHADE
■ One excellent way to get shade over a deck or patio is to train grapevines over an arbor. Four varieties of grapes clamber over the deck arbor *opposite*. The elevated plants create an attractive canopy of cool green leaves on hot summer days. At the same time, fruits will get plenty of ventilation, which will reduce the threat of fungus diseases. As young plants develop, tie canes to posts with soft twine or strips of cloth.

SELECTING FRUIT TREES

Your climate plays a big role in helping you select which fruit trees you can grow. Hardy trees, such as apple, need a period of cold for their growth cycle to continue. But tender trees, such as peach, will get nipped in the bud if winter temperatures are too low.

If you're limited on space, plant a semidwarf (12 to 15 feet tall), dwarf (8 to 10 feet tall), or miniature (6 feet tall) variety. These smaller trees will bear the same size fruit as their standard cousins, at an earlier age.

Some fruit trees are self-fruitful, which means blossoms can be fertilized by pollen from blossoms on the same plant. But other types need a little help from neighboring varieties. To ensure proper pollination of your fruit tree, plant a companion variety within 100 feet, so bees can travel between.

Most apples must be cross-pollinated using simultaneously blooming varieties. Golden Delicious and Rome Beauty will self-pollinate.

Pears generally need two varieties to bear fruit. Kieffer and Duchess are the self-pollinating exceptions. Bartlett and Seckel will not pollinate each other.

Although sour cherries are self-pollinating, sweet cherries—such as Lambert, Bing, Napoleon, and Emperor Francis—need a second variety, such as Black Tartarian, for pollination.

All apricot and nectarine varieties are self-fruitful. Most peach varieties are self-fruitful, but a few require a cross-pollinator. J.H. Hale is one peach that must be cross-pollinated.

Most Japanese (red) plums require a second Japanese plum, but Santa Rosa will pollinate itself, and Burbank and Shiro will not cross-pollinate. European (blue) plums are self-pollinating.

CARE AND MAINTENANCE

To make each year's fruit harvest better than the last, give trees the best possible care throughout the growing season.

WATERING AND FERTILIZING
■ Fruit trees need a lot of water. For best results, dig a trench 6 to 12 inches wide around the plant at the drip line (the farthest tip of branches). Run water from a garden hose into the ditch and let it soak into the ground to the depth of the root system. In young trees, this will be 2 to 3 feet; in older trees, 4 to 5 feet. Use a soil probe to de-

termine what length of time it will take for the water to reach the needed level and how often the soil will dry out and need to be rewatered. Good soil drainage is essential to prevent root damage.

Feed fruit trees with a 5-10-5 fertilizer two to four times between early spring and early summer. If growth is too leafy, lower the dose of fertilizer.

PEST CONTROL
■ Unfortunately, there are a number of insects and diseases that can be a serious threat to the home orchard. Preventive

measures are necessary to keep the trees pest-free. Check with your local extension agent for exact recommendations in your area, but plan on applying a dormant oil spray and a combination fungicide/insecticide spray every eight days from the petal fall stage until one week before harvest. Don't apply insecticides during bloom periods.

THINNING
■ Many fruit trees are too productive for their own good. As a result, limbs droop under the increasing weight until they snap. By removing some of the surplus fruit, you can lighten the load on overladen branches, help reduce disease, and direct more energy to bolster the fruits that remain. If branches still sag, use a pole or stake driven into the ground under the limb for support.

In the fall, wrap fruit trees with a strip of burlap or tree tape to protect tender bark from sunscald, a potential cause of damage or death. Wrapping also discourages rodents from gnawing on the bark.

Inspect the tree about four to six weeks after it blooms, and remove all those fruits that appear diseased, shriveled, or wormy. Later, thin again so fruits will be six to eight inches apart for apples, pears, and peaches, and four inches apart for plums and nectarines. Cherries do not need thinning.

HARVESTING

■ For optimum flavor, harvest plums (*far opposite*) when they part from the stem with a slight twist. Leave plums and peaches on the tree until they're almost ready to eat. Apples are ripe when their seeds are brown; keep them on the tree until they reach their mature color. Pick pears when they're still firm and slightly green; ripen in a cool spot. Harvest cherries when they turn completely red.

1 Fig trees are popular in many warmer climates. In areas colder than Zone 7, they must be gently lifted from the soil and buried in a trench about 6 inches deep.

2 Cover the trench with soil or leaves. In Zone 7, fig trees can be winter-wrapped with burlap or tar paper to prevent frost damage. Uncover them in early spring.

Peaches should be thinned when they are about thumbnail size or slightly larger. To encourage larger fruits, prune early peach varieties 6 to 8 inches apart and late varieties 4 to 5 inches apart.

Birds often will feast on young fruits before harvesttime. To keep birds away, place a "scarecrow" made of beads and small mirrors suspended from the branches with fishing line.

Very fine netting can also be used to discourage birds from feasting on your fruit. The netting can be set in place when the fruit reaches about 1 inch in diameter and left in place until harvest.

CANE AND BUSH FRUITS

Cane and bush fruits are a tasty solution for gardeners who don't have the space for fruit trees. When grown against a fence, berry plants fill otherwise unused space. Before making any selections, ask a local nurseryman which varieties are suitable for your area.

RASPBERRIES
■ Just note the high price of fresh raspberries at the grocery store, and you'll know that this cane fruit is as good as gold. Raspberries were once confined to cold-winter areas, but many newer varieties will thrive in warm regions.

Raspberry varieties fall into two groups: summer-fruiting and everbearing. The summer-fruiting raspberry produces berries on the previous season's canes in early to midsummer. The newer everbearing types bear fruit twice, once in early summer on the previous season's growth, and again in the fall on the current season's growth.

Red and black are the two most common types, but purple and yellow varieties also exist. Red raspberries grow on thorny, erect canes, and black varieties have thorny, arching branches.

BLACKBERRIES
■ Though. frequently confused with black raspberries, blackberries are larger (1 to 2 inches long) and more elongated. Blackberries grow either on 4-foot-tall, erect canes or on trailing vines. Cane varieties thrive everywhere except in extremely cold or extremely warm regions, and vine types are limited to warmer climates. Boysenberries, loganberries, and dewberries are popular trailing varieties. Dewberries bear large, near-black berries; loganberries and boysenberries (*right*) are more red.

GOOSEBERRIES
■ One of the most popular bush fruits is the gooseberry (*opposite*). These tart, ¾-inch berries grow on thorny, 4-foot bushes in cold-winter regions. Berries are green, yellow, pink, or red. Because the gooseberry is an alternate host for white pine blister rust, its cultivation is banned in certain states.

BLUEBERRIES
■ There are several different types of blueberries, but only the high-bush blueberry (*Vaccinium corymbosum*) and

rabbit-eye (*V. ashei*) are recommended for the home garden. Pinkish white flowers bloom in early spring, followed by blue fruit in June. Plant several varieties for cross-pollination.

ELDERBERRIES
■ Elderberries grow in the wild, but cultivated varieties of this hardy bush fruit also are available. Elderberries grow to 10 feet tall and bear purplish black berries in late summer. To ensure fruit production, plant two varieties, such as Adams and Johns.

CARE AND MAINTENANCE OF CANE AND BUSH FRUITS

Proper planting and pruning will bring prolific picking to the berry patch. Once you've planted your cane or bush fruits, your biggest task will be to keep them trimmed. Nothing can become so tangled as a thicket of bramble bushes. Unless bushes are kept under strict control, plants dwindle in yield, ripe fruit becomes inaccessible, and only wildlife can appreciate your planting.

CANE FRUITS
■ Plant cane fruits in early spring as soon as the soil can be worked. Space raspberries and blackberries 30 inches apart, in rows 6 to 8 feet apart. Trailing types need more room; set them 60 inches apart in rows 8 feet apart. Set plants 2 to 3 inches deeper than they

grew before. For support, grow plants between double wires stretched along the length of the row.

After planting, cut back canes to a height of 6 to 8 inches. Mulch with a 4-inch layer of straw to keep the soil around shallow raspberry roots moist and weed-free.

Red raspberries spread by underground runners. These runners are biennial; they grow quickly the first year, produce fruit the second year, then die. In the meantime, new shoots are growing that will produce the following year's crop. Black raspberry stems are also biennial, but they spread by stem-tip plantlets rather than underground runners. New plants form wherever the tip of an old stem touches the ground.

■ **Pruning cane fruits.** Prune cane fruits every year to keep them in fruiting condition. Because canes bear only once, remove old canes after harvest. Cut back all but five to eight of the strongest year-old stems to produce fruit the following year. In early spring, shorten black raspberry canes to 6 inches and others to 12 inches; shorten red raspberries to 24 to 36 inches. Do not prune everbearing types in the spring. In midsummer, pinch back trailing varieties to encourage lateral branching.

BUSH FRUITS
■ **Gooseberries and currants.** Plant gooseberries and currants in spring or fall (fall is best). Set plants deeper than they grew before, spacing them 4 to 5 feet apart. After planting, cut branches back to 12 inches. Prune out old wood and low-hanging branches each year while shrubs are dormant.
■ **Blueberries.** Plant blueberries in spring or fall, 4 to 7 feet apart, in rows 8 feet apart. Buy only two- to three-year-old plants, and set them at the level they grew previously. After planting, cut tops back to 6 to 10 inches. Mulch heavily to protect shallow roots.

Fertilize blueberries with ammonium sulfate each spring before the buds open. Prune out old and weak stems every winter after the plants are established. Protect berries from birds by using nets, cheesecloth, or chicken wire.
■ **Elderberries.** Space your elderberry plants 6 feet apart in rows 10 feet apart. If pruning is required, do it during the winter before new growth starts. Cut back older stems occasionally to stimulate new growth and heavier bearing.

Remove old canes (those that bore fruit) and deadwood after fall harvest. Shorten remaining canes, then thin out all but 5 to 8 of the strongest year-old canes.

When growth gets too long or ungainly, cut it back. If red raspberries produce two crops, a light crop will first form in fall. Cut canes back for spring berries.

Control weeds and keep the soil moist with a 4-inch-thick mulch of straw. Tender loganberries and boysenberries need mulch for winter protection.

For top-notch fruiting, train red raspberries on a trellis made of wire stretched along each side of a 12-inch-wide row. Trailing and semi-trailing blackberries also perform well when trellised.

Save garden space by planting a row of raspberry plants along a property-line fence. Fruit will be supported upright for easy harvesting, but pruning will take more care and time.

Birds love blueberries and will eat the entire crop if it is not protected. Along with a protective covering of netting, cloth, or wire, use something reflective, such as strips of aluminum foil or a pie tin.

GRAPES

Grape vines are perfect candidates for the small-lot garden. Use them on a trellis for privacy, or let them ramble over an arbor (*opposite*) for shade.

Plant two varieties to ensure cross-pollination. Most varieties are self-fertile but will produce larger clusters of fruit if interplanted. For an all-season crop, combine early-bearing varieties like Fredonia, Seedless, and Delaware with the mid- to late-season varieties; such as Niagara, Concord, Golden Muscat, Catawba, and Vinered.

HOW TO PLANT
■ Plant grapes in fall or in early spring as soon as the soil can be worked. Choose a sunny location that has rich, well-drained soil. Good air circulation also is necessary to prevent mildew, rot, and damage from spring frost. If possi-

ble, run the vines in an east/west orientation to take full advantage of sunlight.

Before planting new stock, clip back roots to 8 or 12 inches long and vines to the second set of buds. Spread the roots out evenly in the planting hole. Space plants at least 8 feet apart. Side-dress with well-rotted manure or all-purpose fertilizer, and mulch with 2 to 4 inches of leaves or straw.

HOW TO PRUNE
■ At planting time, prune each vine to the best single cane, and prune that cane back to 2 or 3 buds. In following years, prune grapes in early spring while the vines are still dormant. If you want shade for a patio or screening for privacy, limit pruning to the removal of scraggly growth and old canes. Let a few new canes develop each year for a

small harvest. Old canes produce only shade. A good grape arbor should provide plenty of ventilation around the canes to discourage mildew diseases.

If you grow grapes on a 2-wire trellis, such as in the drawing *below,* train the original cane to grow straight up a stake or string to the top wire. Cut the cane off just above the top wire to encourage arms. Train arms during the second growing season; all other shoots, suckers, and canes should be removed.

FERTILIZING
■ After spring planting, feed each vine with 4 ounces of 5-10-5 fertilizer, applying it no closer than 12 inches from the base of the trunk. In following years, feed as early in the spring as possible. Apply a pound and a half of fertilizer per mature vine.

Although most grapes are self-pollinating, you will get better results if you plant two varieties in your garden. Clusters of fruit will be larger. Harvest grapes when they're sweet and juicy.

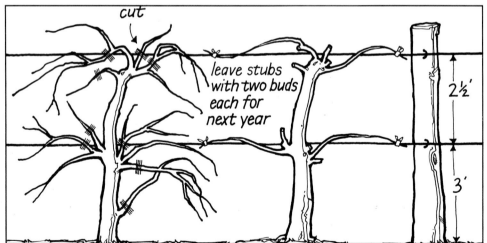

A good pruning method, called the four-arm "Kniffen system," is shown *above.* During the plant's second year, four arms are allowed to develop, two in each direction from the trunk, one set per wire.

Select four other canes, one from each of the original arms, and prune to two buds. These spurs will form new canes. The next spring, leave four canes and four spurs; remove all other wood.

STRAWBERRIES

Nothing beats the flavor of garden-ripe strawberries, freshly plucked from the backyard patch. Because they give luscious rewards without a lot of work, strawberries provide a great starting point for beginning fruit growers. Tuck single plants into a vegetable garden, or mass several together to make an ornamental and edible ground cover. Or, if your garden is limited to patio space, grow strawberries in a strawberry jar (*opposite*). Ask a horticulturist at your local nursery about what varieties are suited for your region.

Plant sets in any sunny, well-drained garden location. To save money, you also can grow strawberries from seed. Sow seeds indoors six to eight weeks before outdoor planting. For a bumper crop the second year after planting, remove first-year runners and blossoms as soon as they appear.

TYPES OF STRAWBERRIES

■ **Everbearing.** Their name may imply continuous harvests, but everbearing strawberries actually produce two separate crops each year. Their first crop ripens in June, followed by a heavier second crop of smaller berries in early fall. Because they require a long growing season, everbearing varieties are not always recommended for northern regions. Popular varieties include Ogallala, Dunlap, Geneva, Ozark Beauty, and Superfection.

■ **June-bearing.** June-bearers produce fruit in early summer. Although plants offer a bigger summer yield than everbearing types do, they stop fruit production after the first harvest. In the June-bearing group, try Earliglow, Fairfax, Surecrop, or Sparkle.

■ **Day-neutral.** Newer day-neutral varieties produce berries all summer long. Unlike conventional strawberry plants, which bear only in the cool days of early June (or, in the case of everbearing, in June and again in the fall), day-neutral strawberries produce their flowers and fruit regardless of day length. That means you can serve up fresh strawberry shortcake in the winter, too, if you move your plants indoors and tend them carefully. Tristar, Tribute, and Brighton are three recommended day-neutral varieties.

■ **Alpine.** Unlike their cultivated cousins, alpine strawberries do not send out runners. Instead, they form compact mounds that are perfect for edgings around vegetable, flower, or herb gardens. Fruits are small and very sweet, much like the strawberries you find growing wild. Each plant will bloom for many years. Try Alexandria or Ruegen Improved for best results.

CARE AND MAINTENANCE

Strawberry plants will set fruit their first season, but they'll be more productive if you make them wait a year. For maximum results, plant strawberries in well-prepared soil and pinch off all blossoms the first season. That way, plants can store up berry-making power to make a jumbo harvest the following year.

SOIL PREPARATION
■ Strawberries require a well-drained, weed-free planting bed. If you're breaking ground for the first time, consider planting a vegetable crop in that space the first year. That way, the soil will be in better condition and rid of stubborn weeds by the second year. Before planting strawberries, spade in plenty of compost, and apply a complete fertilizer, such as 5-10-5.

GROWING METHODS
■ Proper planting depth should get top priority when you set out new strawberry transplants. Match the new soil line with the depth that the plant grew at the nursery. Make sure roots are completely underground, but avoid covering the crown. Mulch will help keep fruit clean, conserve moisture, and suppress weeds.

The technique you use for growing strawberries will depend on the size of berry and yield you want. Close, dense planting results in heavy yields of smaller berries, and open, well-spaced planting offers a lower yield but larger berries. The chief difference between each system is what you do with the runners, the baby plants that develop from the main plant.

■ **Hill planting.** For large berries and a neatly manicured strawberry bed, set plants 12 inches apart in rows spaced about 18 inches apart. Remove all run-

ners as they develop. This method results in healthy, strong central plants that bear maximum-size berries.

■ **Matted row.** For the least maintenance, use the matted-row planting method. Set transplants 18 to 24 inches apart, and after the first growing season, allow runners to root where they land. The result will be an informal mass instead of tidy rows. Remove first-year blossoms to strengthen plants. You'll get a bumper crop of smaller berries the following year.

REJUVENATING AN OLD BED
■ A well-cared-for strawberry patch can last up to five years. To renew a matted-row bed, cut plants back with a rotary mower set at a 4-inch cutting height. Feed with a 5-10-5 fertilizer and water. To prevent disease and insect problems, replace plants after five years.

Encourage runners (no more than six per mother plant) to form new plants every 9 inches. Pin down the runners with U-shaped pins or bury stems with soil. Remove unwanted runners.

1 Buy disease-free strawberry transplants from a reputable nursery. Spade soil to a depth of 8 to 10 inches, and mix in compost or manure to improve drainage.

2 Set plants at the same depth they grew at the nursery, with the crown planted just above the soil line. Spread the roots evenly over the soil mound.

3 Cover the roots with soil and gently firm into place. As you work, protect all unplanted strawberries by covering them with wet burlap or newspaper.

For large yields, use the matted-row planting method. Set the transplants 18 to 24 inches apart, and allow runners to root where they land. The result will be a solid mass instead of tidy rows.

To get jumbo-size berries, follow the hill planting method. Set plants 12 inches apart in rows spaced 18 inches apart. Pinch off all runners when they appear, to encourage a strong central plant.

Give strawberry plants a winter mulch after the first fall frost. In spring, loosen the mulch as plants turn green, but don't remove it completely until temperatures are dependably warm.

CONSIDER YOUR CLIMATE

The climatic conditions in your area are a mixture of different weather patterns: sun, snow, rain, wind, and humidity. A good gardener is aware of all of the variations in temperature and conditions in his or her own garden, from how much rainfall it receives each year to the high and low temperatures of a typical growing season.

The zone map at *right* gives an approximate range of minimum temperatures across the country. Most plants are rated by these zones for conditions where they grow best.

However, zone boundary lines are not absolute. You can obtain the general information for your area from your state agricultural school or your county extension agent.

Be sure to study the microclimates that characterize your own plot of ground. Land on the south side of your house is bound to be warmer than a constantly shaded area exposed to cold, northwest winds. Being aware of the variations in your garden will help you choose the best plant for the prevailing conditions and avoid disappointment.

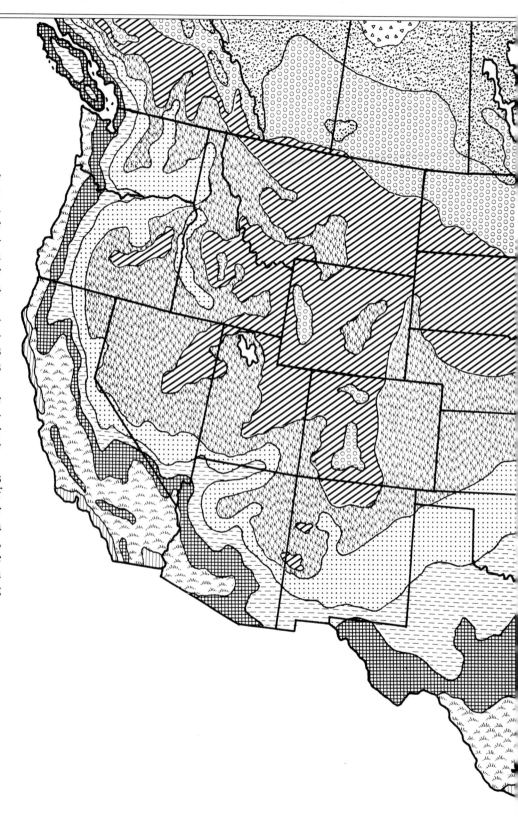

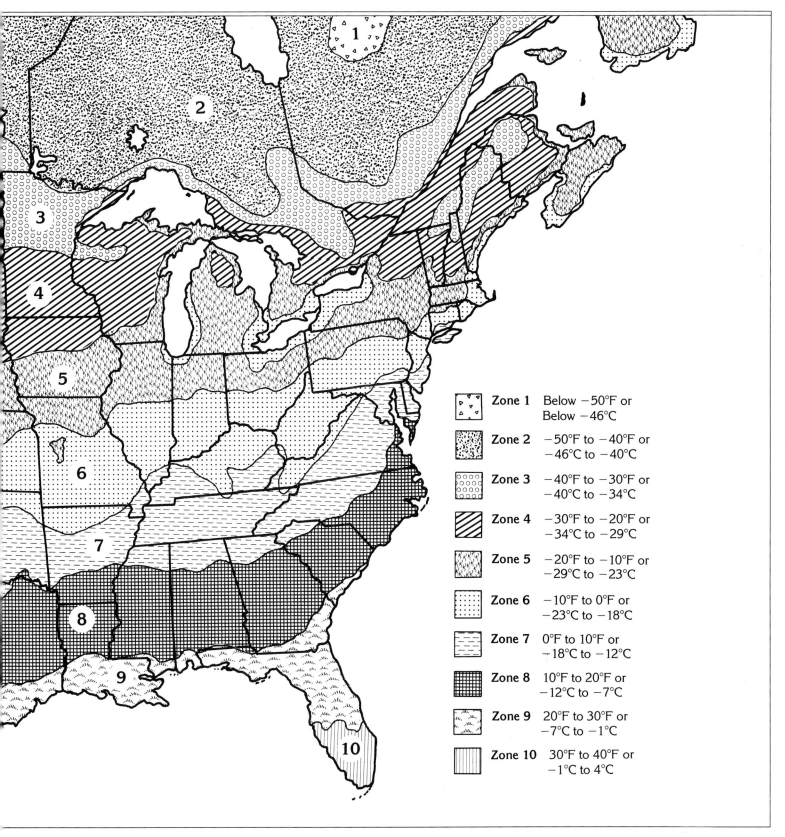

Zone 1 — Below −50°F or Below −46°C

Zone 2 — −50°F to −40°F or −46°C to −40°C

Zone 3 — −40°F to −30°F or −40°C to −34°C

Zone 4 — −30°F to −20°F or −34°C to −29°C

Zone 5 — −20°F to −10°F or −29°C to −23°C

Zone 6 — −10°F to 0°F or −23°C to −18°C

Zone 7 — 0°F to 10°F or −18°C to −12°C

Zone 8 — 10°F to 20°F or −12°C to −7°C

Zone 9 — 20°F to 30°F or −7°C to −1°C

Zone 10 — 30°F to 40°F or −1°C to 4°C

INDEX